PHYSICAL SCIENCE

FORMS OF ENERGY

By Christina Earley

A Stingray Book

SEAHORSE PUBLISHING

Teaching Tips for Caregivers and Teachers:

This Hi-Lo book features high-interest subject matter that will appeal to all readers in intermediate and middle school grades. It may be enjoyed by students reading at or above grade level as well as by those who are looking for age-appropriate themes matched with a less challenging reading level. Hi-Lo books are ideal for ELL readers, too.

Each book appeals to a striving reader's age and maturity level. Opportunities are provided for students to read words they already know while encountering a limited number of new, high-interest vocabulary words. With these supports in place, students will read more fluently while increasing reading comprehension. Use the following suggestions to help students grow as readers.

- Encourage the student to read independently at home.
- Encourage the student to practice reading aloud.
- Encourage activities that require reading.
- Establish a regular reading time.
- Have the student write questions about what they read.

Teaching Tips for Teachers:

Before Reading

- Ask, "What do I know about this topic?"
- Ask, "What do I want to learn about this topic?"

During Reading

- Ask, "What is the author trying to teach me?"
- Ask, "How is this like something I already know?"

After Reading

- Discuss how the text features (headings, index, etc.) help with understanding the topic.
- Ask, "What interesting or fun fact did you learn?"

TABLE OF CONTENTS

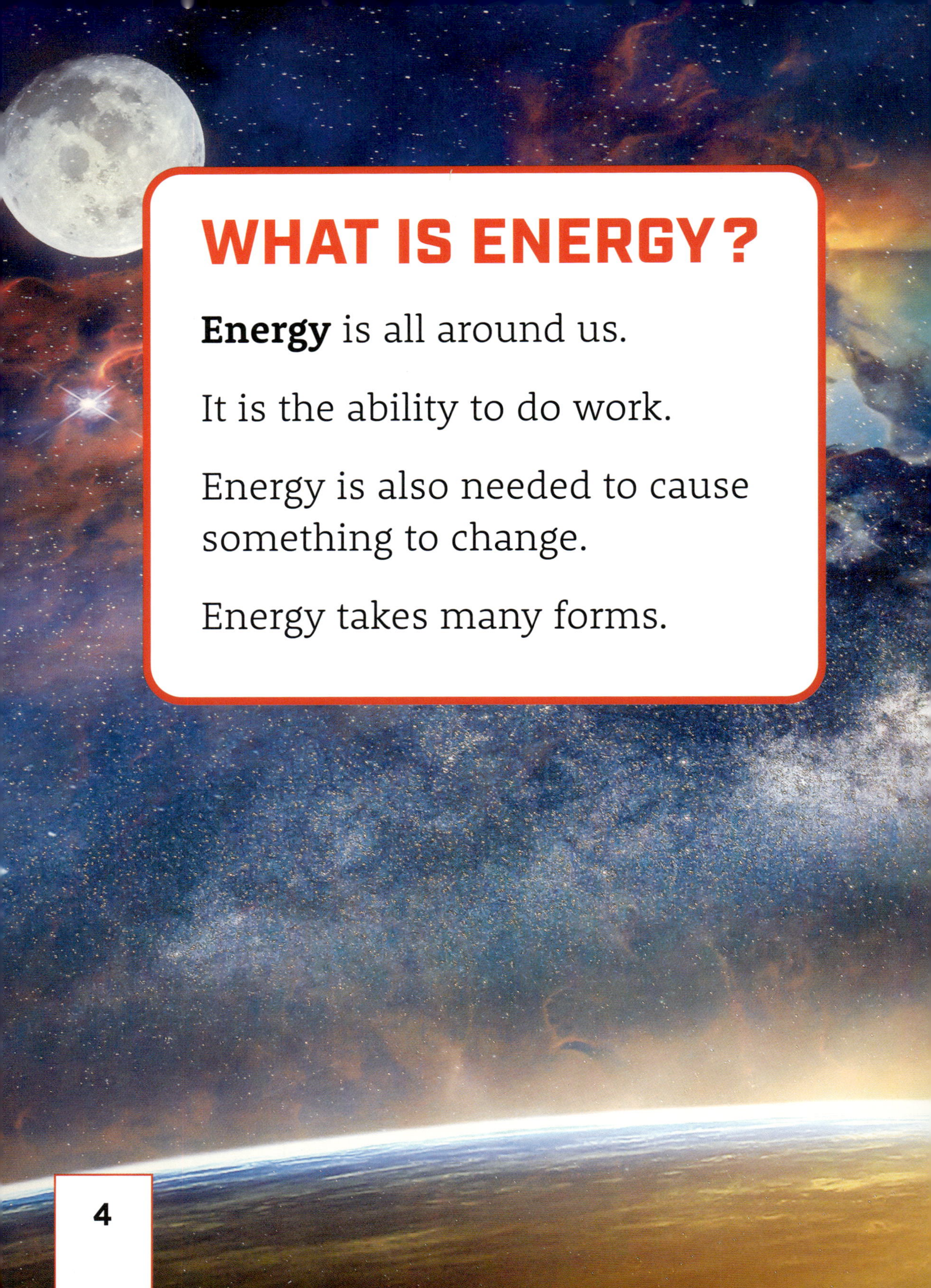

WHAT IS ENERGY?

Energy is all around us.

It is the ability to do work.

Energy is also needed to cause something to change.

Energy takes many forms.

FUN FACTS

Energy can be stored, or potential. It can be working, or kinetic.

MECHANICAL ENERGY

Mechanical energy is motion or movement energy.

It is all the energy an **object** has because of its motion and position.

A car uses mechanical energy to move people from place to place.

THE LION KING
music

LIGHT ENERGY

Light energy is radiant energy.

It is a type of energy that travels in waves.

The sun is one kind of light energy.

Using a flashlight to read is another use of light energy.

FUN FACTS

It takes 8 minutes 17 seconds for light to travel from the Sun's surface to Earth.

THERMAL ENERGY

Thermal energy is heat energy.

It is the amount of heat that can be measured in an object.

Baking a pizza uses thermal energy.

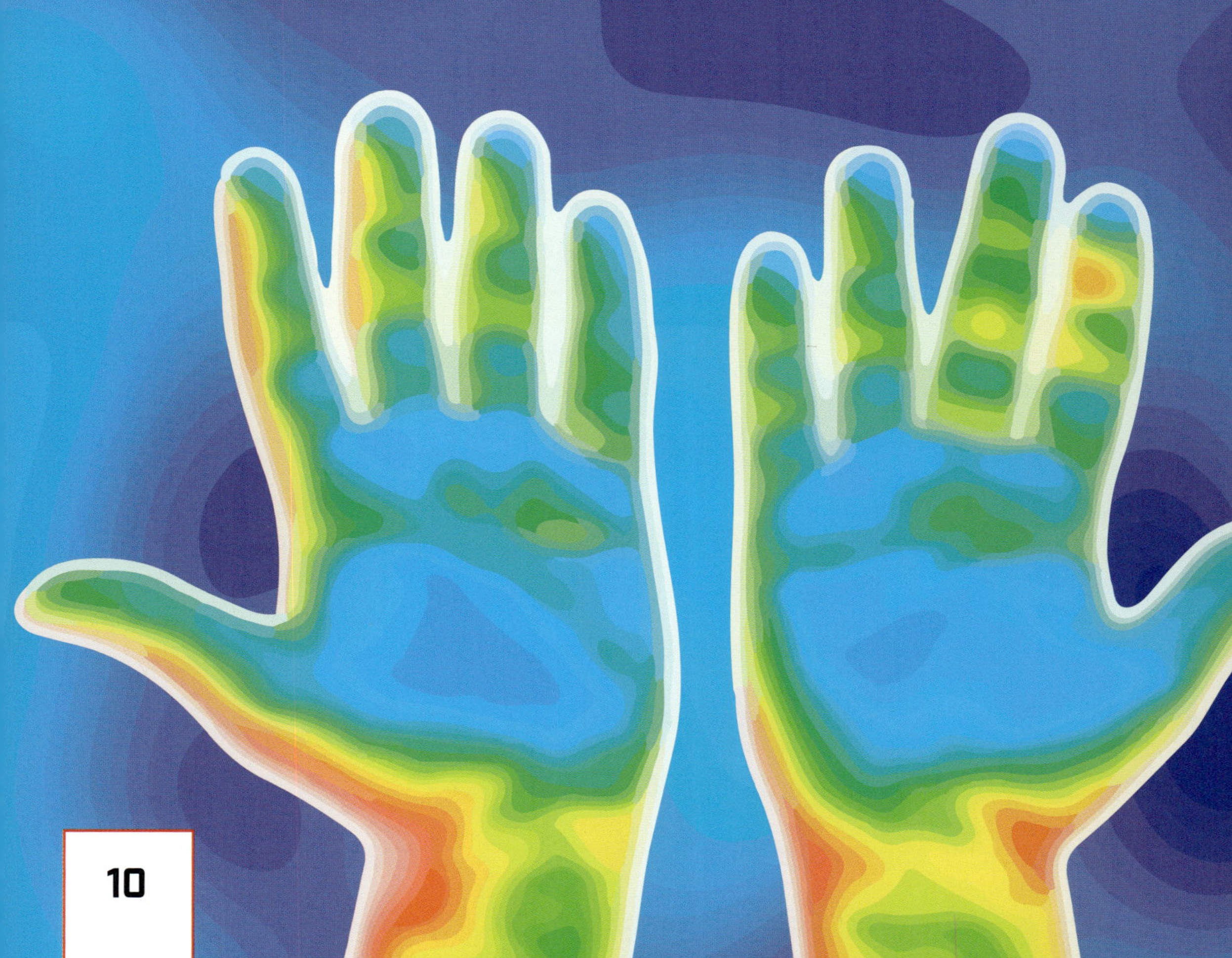

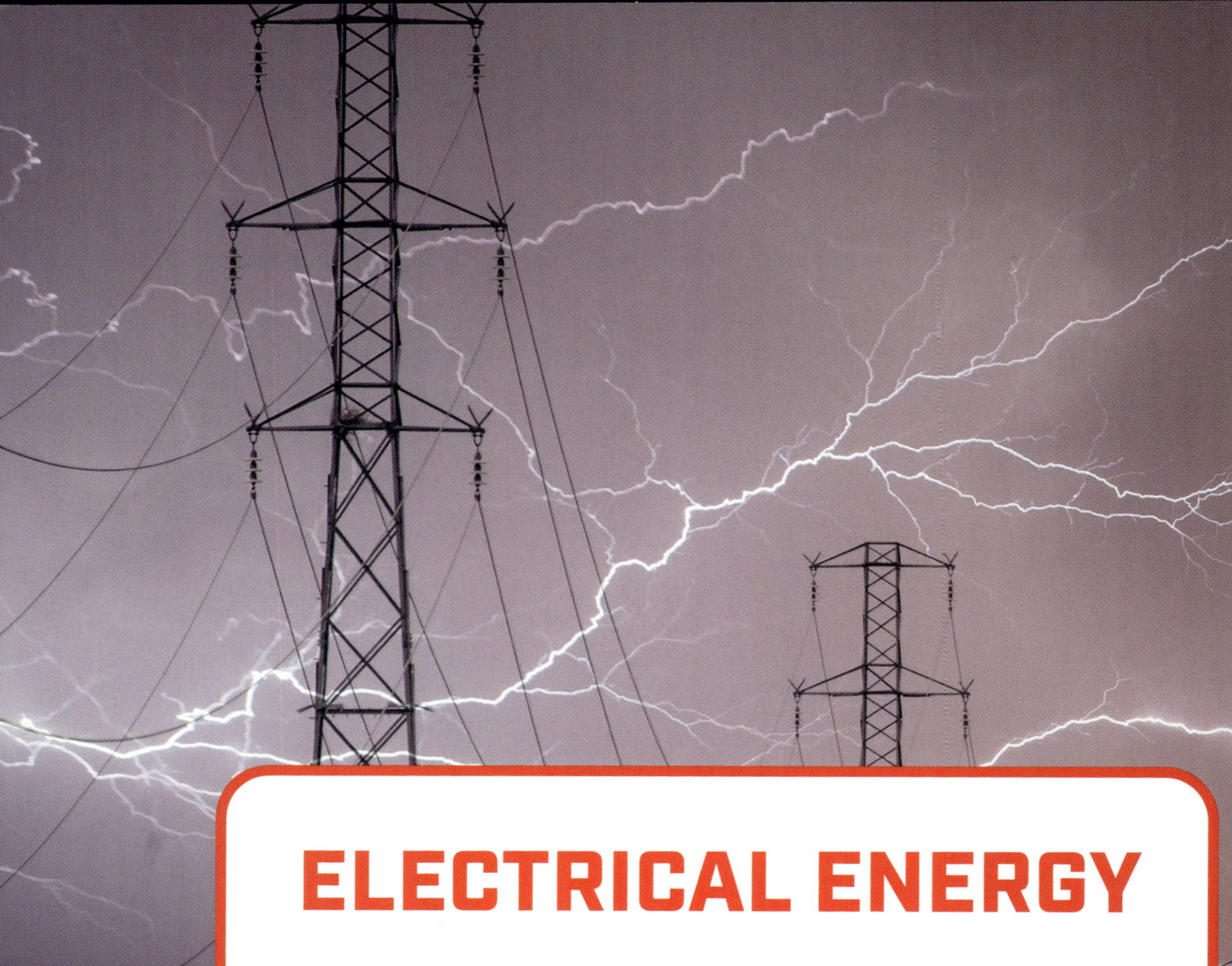

ELECTRICAL ENERGY

Electricity is electrical energy.

It is the movement of an electrical charge from one place to another.

The **power** of an atom's charged **particles** force an action or move an object.

FUN FACTS

Electricity travels at the speed of light.

SOUND ENERGY

Sound energy is made from vibrations you can hear.

These vibrations make waves.

Music is sound energy.

FUN FACTS

There is no sound in space because there is nothing with atoms for the sound to travel through.

CHEMICAL ENERGY

Chemical energy comes from how atoms and **molecules** interact.

This makes a chemical reaction.

The gas in a car gives chemical energy to make it move.

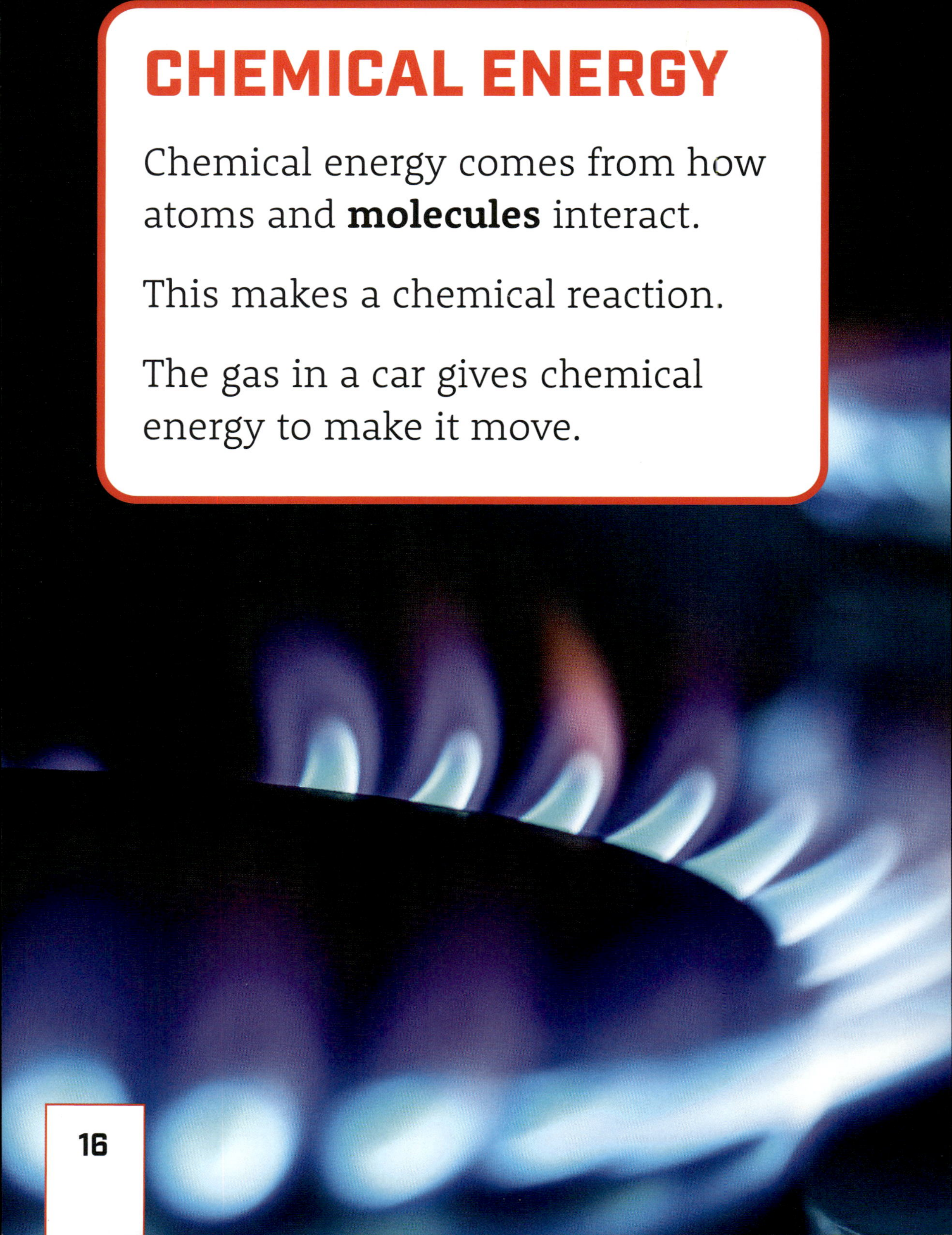

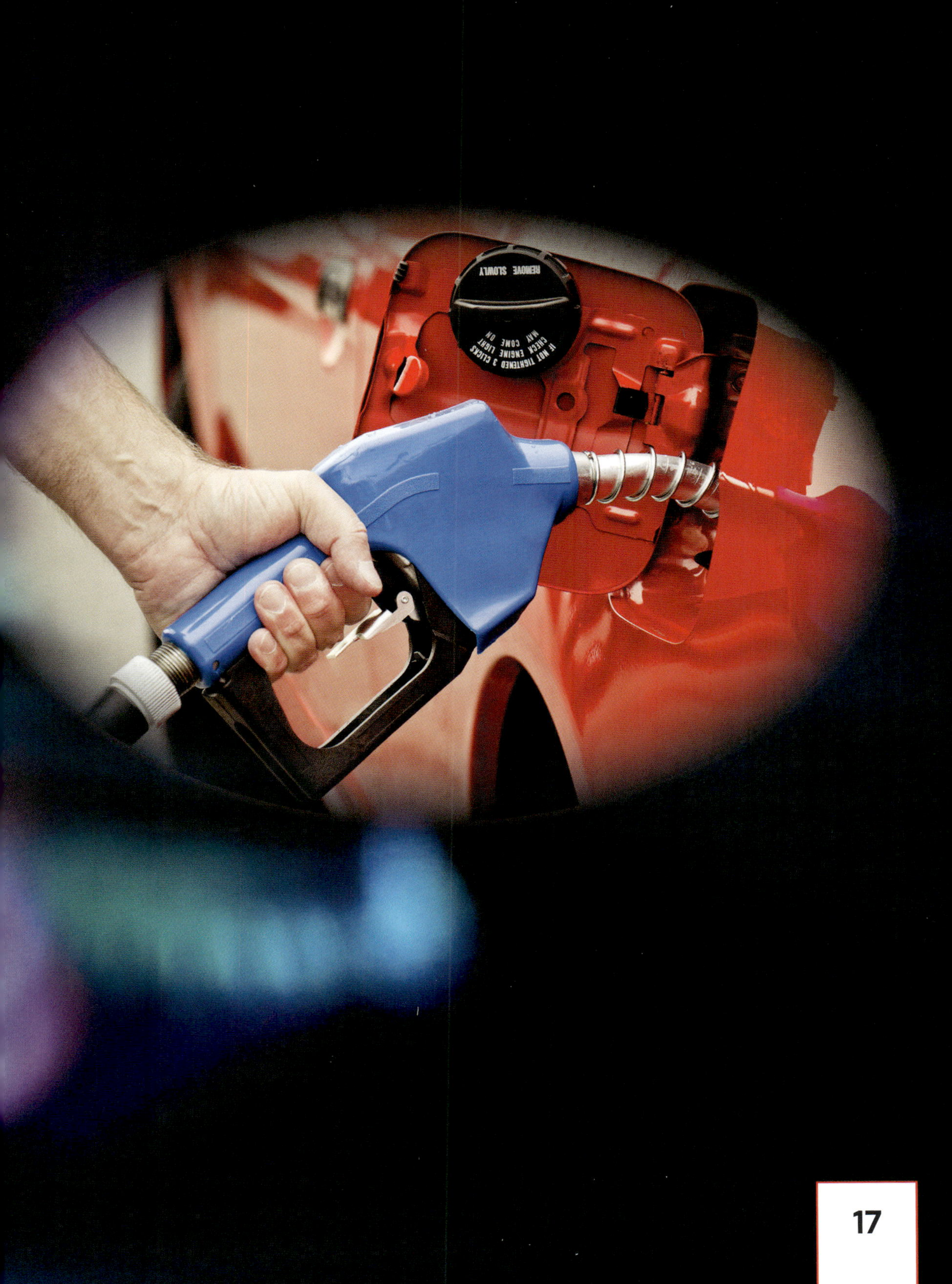
REMOVE SLOWLY
IF NOT TIGHTENED 3 CLICKS CHECK ENGINE LIGHT MAY COME ON

CAREER: SOLAR ENERGY ENGINEER

A solar energy engineer designs a solar panel power system for a home or business.

The Sun's energy is changed into electrical energy.

This energy can be used to heat a swimming pool.

This energy can be used to give power to a building.

INVESTIGATE: SOLAR BALLOONS

Materials:

- Six empty plastic bottles of the same size (8 oz. – 16 oz.)
- Poster paint: black, white, yellow, red, green, blue
- Small balloons

Procedure:

(1) Paint the outside of each bottle a different color.

(2) Place one balloon on the mouth of each bottle. Make sure there is a tight seal.

(3) Put the bottles in the bright sunlight.

(4) Observe the bottles. Which ones have expanded? Which ones have stayed the same?

(5) Feel the bottles. Which ones are hot? Which ones are cool?

(6) Compare the temperature of the bottles with the balloons.

THE SCIENTIFIC METHOD

- Ask a question.
- Gather information and observe.
- Make a **hypothesis** or guess the answer.
- Experiment and test your hypothesis, or guess.
- Analyze your test results.
- Modify your hypothesis, if necessary.
- Make a conclusion.

SCIENTIST SPOTLIGHT

Anna Stork and Andrea Sreshta are the inventors of the LuminAID solar light. At graduate school in 2010, they were assigned to create a way to help with disaster relief from an earthquake in Haiti. The pair designed a light that is **inflatable**, waterproof, and solar-powered.

Over the past ten years, they have continued to improve the lantern. They send the lanterns to disaster sites. People who enjoy the outdoors also are purchasing these lanterns. Stork and Sreshta are giving light to the world in the darkest of times.

GLOSSARY

energy (EN·er·jee): quantitative property that can be transferred to a body or machine to make it do work

hypothesis (high·POTH·uh·sis): proposed explanation for a phenomenon

inflatable (in·FLAY·tuh·buhl): able to be filled with air

molecules (MOL·uh·kyoolz): groups of atoms bonded together, representing the smallest fundamental unit of a chemical compound

object (OB·jikt): a thing that can be seen or touched

particles (PAHR·ti·kuhlz): minute portions of matter

power (PAU·er): a source or means of supplying energy

INDEX

AFTER READING QUESTIONS

1. What is thermal energy?
2. Using a flashlight to read is an example of what type of energy?
3. What type of energy does a car use while driving?

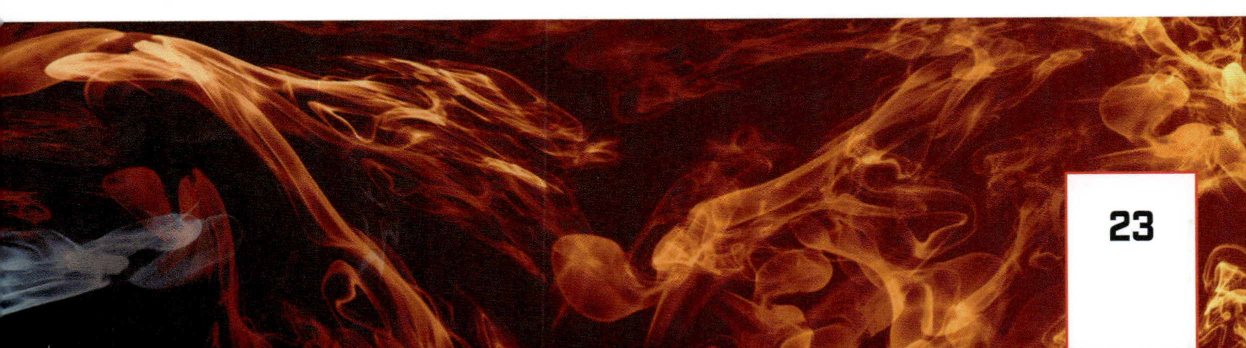

About the Author

Christina Earley lives in South Florida with her husband, son, and dog. Her favorite subject in school was science. She enjoys learning the science behind the world around her by asking questions such as how do roller coasters work. She loves mint chocolate chip ice cream and mermaids.

Written by: Christina Earley
Design by: Kathy Walsh
Editor: Kim Thompson

Photographs/Shutterstock: Cover ©Open Studio: Cover, Pg 1, 3 © Krzysztof Bubel: Pg 4 ©Nuttawut Uttamaharad: Pg 6 ©Have a nice day Photo: Pg 8 ©New Africa, ©Guenter Albers: Pg 10 ©cipta studio: Pg 11 ©Varavin88: Pg 12©muratart: Pg 13 ©Animatic: Pg 14 ©ioat: Pg 15©Vershinin89: Pg 16 ©Denis Torkhov: Pg 17 ©Carolyn Franks: Pg 18 ©Fit Ztudio: Pg 19 ©VAKS-Stock Agency: Pg 20 ©santima studio: Pg 21 ©Pixel-Shot: Pg 22 ©kkssr

Library of Congress PCN Data
Forms of Energy / Christina Earley
Physical Science
ISBN 978-1-63897-115-3 (hard cover)
ISBN 978-1-63897-201-3 (paperback)
ISBN 978-1-63897-287-7 (EPUB)
ISBN 978-1-63897-373-7 (eBook)
Library of Congress Control Number: 2021945216

Printed in the United States of America.

Seahorse Publishing Company
www.seahorsepub.com 1-800-387-7650

Published in the United States
Seahorse Publishing
PO Box 771325
Coral Springs, FL 33077